Food 226

太空美食

Space Food

Gunter Pauli

[比] 冈特·鲍利 著

[哥伦] 凯瑟琳娜·巴赫 绘

颜莹莹 译

上海远东出版社

丛书编委会

主　任：贾　峰

副主任：何家振　闫世东　郑立明

委　员：李原原　祝真旭　牛玲娟　梁雅丽　任泽林

　　　　王　岢　陈　卫　郑循如　吴建民　彭　勇

　　　　王梦雨　戴　虹　靳增江　孟　蝶　崔晓晓

特别感谢以下热心人士对童书工作的支持：

匡志强　方　芳　宋小华　解　东　厉　云　李　婧

刘　丹　熊彩虹　罗淑怡　旷　婉　杨　荣　刘学振

何圣霖　王必斗　潘林平　熊志强　廖清州　谭燕宁

王　征　白　纯　张林霞　寿颖慧　罗　佳　傅　俊

胡海朋　白永喆　韦小宏　李　杰　欧　亮

目录

Contents

ZURI Learning Initiative

两个年轻女孩正仰望天空，计划着她们的火星之旅。她们确信她们的梦想之旅很快就能实现，尽管其他人觉得她们是活在幻想世界里的梦想家。

　　"我并不担心如何发动宇宙飞船，"安妮边说边摇晃着她金色的卷发，"这个问题已经解决了。"

Two young girls are looking up at the sky, planning their trip to Mars. They are sure that their trip of a lifetime will soon become a reality, despite others considering them dreamers, living in a fantasy world.

"I am not worried about how we will power our spaceship," Anne says, shaking her blonde curls. "That issue has been resolved."

计划着她们的火星之旅……

Planning their trip to Mars …

配置3D打印机来制造......

set up 3-D printers to manufacture ...

"我同意，"凯瑟琳回答，"我们需要原材料，所以我们要开采小行星，配置3D打印机来制造我们需要的任何东西。解决了！"

"真正的问题是食物。我们要吃什么，喜欢吃什么？这才是我们9个月火星之旅计划有趣的地方。"

"Agreed," Katherine responds. "We need materials, so we'll mine asteroids and set up 3-D printers to manufacture whatever we want. Solved!"

"The real question is food. What will we eat, and what will we enjoy eating? This is the fun part of planning our nine-month-long trip to Mars."

"我们考虑一下到达火星所需的这九个月之后吧。我们踏上火星后的头几年吃什么呢？"

"如果我们被限制在一个狭小的空间里，很快就会感到无聊，最后会像老虎一样——在笼子里踱步，紧张、抑郁，行为怪异。"

"Let's think beyond the nine months it takes to get there. What will we eat once we arrive, and for the first few years after we have set foot on Mars?"

"If we are confined to a tiny space, we'll soon get bored and end up like a tiger – pacing its cage, stressed, depressed and behaving strangely."

最后会像老虎一样——在笼子里踱步。

end up like a tiger - pacing its cage.

我担心会变得无聊。

I am worried about getting bored.

"和你一样，我不太担心食物和我们能制造的东西，我担心会变得无聊。这将是我们要克服的最大挑战。"

　　"要知道，这不是什么新鲜事。在医院或监狱待久了，或在隔离期间孤独无伴的人也面临同样的风险。"

"Like you, I am not too worried about food and the things we can make, I am worried about getting bored. That will be the greatest challenge for us to overcome."
"That's nothing new, you know. People who spend a long time in hospital or prison, or endure isolation during quarantine, run the same risk."

"那我们应该设想一些全新的活动和产品，而不是尝试种植我们在地球上吃了这么多年的东西。"

　　"嗯，不错的主意，"安妮回答，"我们不应该做思乡心切的地球人，我们应该让自己成为有干劲和满足感的太空人。"

"Let's then imagine some completely new activities and products, instead of trying to farm what we have been eating on Earth through the ages."

"Hmmm, I like that," Anne replies. "We should not leave for Mars as Earthlings who will become homesick. We should reinvent ourselves as inspired and satisfied Spacelings."

太空人

Spacelings

"完全正确！比如可以与机器人共舞，或者做一些无重力的雕塑？"

"还可以用无重力乐器代替风笛来演奏音乐。"

"到底什么是无重力乐器？"凯瑟琳问道。

"一种不使用重力的乐器。"

"Exactly! Doing things like inventing a dance with robots, or making sculptures in forms that are gravity-free?"

"And playing music on a telemetron, instead of bagpipes."

"What on earth is a telemetron?" Katherine wants to know.

"An instrument that uses zero gravity."

"从来没见过，也没有听说过。但是，你知道，我的创造力更倾向于食物。要解决食物的问题，并不像一次有挑战性的露营那么简单。"

"艺术、音乐、运动，还有美食！当我们离开地球，踏上通往火星的单程旅行时，这将使我们不再无聊。"

"这意味着我们必须重新考虑一下厨房。"

"Never seen or, at least, heard that. But, you know, my creative mind gravitates more towards food. We cannot treat the issue of food as if this is simply a challenging camping trip…"

"Art, music, movement, and great food! That is what will make us thrive as we leave Earth behind us, on this one-way ticket to Mars."

"And that means we will also have to rethink the kitchen."

通往火星的单程旅行

One-way ticket to Mars

360度漂浮着

served floating in 360 degrees

"不仅如此，我还想设计一个全新的餐厅。食物在我们的日常生活中非常重要。我们组建团队，分享美好回忆，我们还需要一个消遣的地方！"

"我在想象这种感觉：一个柠檬球里的姜汁泡泡球，突然跑到甜菜汁泡泡球内，它们360度漂浮着。盘子、刀叉都不需要了，直接坐在桌前即可。"

"More that just that, I want to design an entirely new dining room. Food will be so important in our daily routine, as we build teams, sharing fond memories. We'll also need a place to hang out!"

"I'm dreaming of this sensation: a bubble of ginger, inside a lemon sphere, popping into a beetroot juice globule, served floating in 360 degrees. Gone are the plates, knives and forks, and sitting up straight at the table."

"是啊！现在我们已经踏上了真正的创意烹饪之路——品尝我们在地球上永远无法尝到的美食。当我们在家里的时候，你觉得爸妈会同意我们表达自己的风格吗？"

　　……这仅仅是开始！……

"Yes! Now we are on track for some real creative cooking – savouring things we never could on Earth. Do you think Mom and Dad would approve of us expressing our style while we are still at home?"

... AND IT HAS ONLY JUST BEGUN!...

...... 这仅仅是开始！......

... AND IT HAS ONLY JUST BEGUN! ...

Did You Know ?

你知道吗？

Staff of offshore oil rigs, supertankers, submarines and Antarctic research stations are served great food on cloth-covered tables to maintain morale and productivity in isolated, remote, confined situations.

海上石油钻井平台、超级油轮、潜艇和南极科考站的工作人员会在铺着桌布的桌子上享用丰盛的食物，以便在与世隔绝、偏远、封闭的环境中保持士气和生产力。

1 kg $$$

Every kilo of weight a spaceship transports in space costs thousands of dollars. This implies that food must be lightweight and compact. It also has to last a long time.

宇宙飞船在太空中运载的每一千克物品都要花费数千美元。这意味着食物必须是轻质和压缩的，还必须有很长的保质期。

On Earth, crumbs fall; in microgravity, they can end up anywhere, including inside critical equipment or astronauts' lungs. Food in space comes in the form of squeezable purées and "intermediate moisture bites".

在地球上，面包屑会落在地上；在微重力下，它们会落在任何地方，包括关键设备内部或宇航员的肺部。太空中的食物呈现为可挤压的糊状和"中等湿度口感"。

Space cutlery has been reduced to a pair of scissors, for opening packages, and a spoon, for scooping out their contents. Cooking is simplified to rehydrating food by adding hot water.

太空餐具已经简化为一把剪刀（用来打开包装），以及一把勺子（用来舀出里面的东西）。烹饪简化为加入热水使食物再水化。

For short lunar trips, space travellers are happy with snacks and treats. For long trips to Mars, people will need food with a variety of flavours, textures and temperatures.

对于短途的月球旅行，太空旅行者很喜欢吃零食和大餐。在前往火星的长途旅行中，人们会需要各式各样口感、风味和温度的食物。

Liquids in space behave like wobbly blobs rather than a fluid or droplets. Spheres of water with ginger and lemon encapsulated into spherical blobs would not be served on a plate but garnished in 360°.

液体在太空中更像不稳定的水团，而不是水流或水滴。封在球状水团中的生姜和柠檬提取物不会摆在盘子里，而是360度全方位摆放。

食物的味道是微生物消化的副产品。在微重力环境下，微生物和人类的新陈代谢方式不同。由此产生的味道可能与我们在地球上所期待的不同。

Food tastes are byproducts of microbial digestion. Metabolism works differently in microgravity, for microbes as well as for humans. The resulting flavours may differ from what we expected while on Earth.

如果宇宙飞船失去气压，氧气面罩不会自动掉落；相反，每位宇航员不得不通过安装在中央通道和墙壁上的氧气盒来保持呼吸。

Should a spaceship lose pressure, oxygen masks will not drop automatically; instead, each astronaut will have to make his or her way over to the oxygen boxes mounted along the centre aisle and walls, to resume breathing.

What about a dinner without plates, forks, knives, or spoons?

没有盘子、叉子、刀、勺子的晚餐会是什么样的?

Is it possible to play bagpipes in Space?

有可能在太空吹奏风笛吗?

How difficult would it be to mine meteorites?

开采陨石有多困难?

Would you bother to redesign the kitchen and the dining room to accommodate zero gravity?

你会费心重新设计厨房和餐厅以适应零重力环境吗?

Do It Yourself!

自己动手！

Research the food that you would like to have on your space trip to Mars. Do you expect to be able to eat what you eat at home? You will most likely have to change your diet completely. Are you ready for that? Look into how early emigrants who undertook long sea voyages, over 500 years ago secured food for the long trip. On a 9-month-long trip to Mars, would you be able to farm everything on the way there? Ask your friends and family members who would want to travel to Mars with you, and what their ideas are about solving the issue of food on the way, and once they arrive. Now discuss the best solutions.

研究一下你在去火星的太空旅行中想要吃的食物。你会希望吃到在家里吃的食物吗？你将很可能彻底改变饮食。你做好准备了吗？调查一下500多年前长途航海的早期移民是如何为长途旅行获取食物的。在为期9个月的火星之旅中，你能在去火星的路上种植所有的东西吗？问问你的朋友和家人谁想和你一起去火星旅行，问问他们对解决途中和到达后的食物问题有什么想法。讨论一个最好的解决方案。

学科知识
Academic Knowledge

生物学	在船上种植食物只能提供日常饮食需求的一小部分。
化 学	地球上的大气主要是氮气和氧气,火星大气的95%是二氧化碳,2.7%是氮气,只有0.13%的氧气;火星上有由二氧化碳构成的雪花;在火星上,二氧化碳和水可以产生液态甲烷;火星上的水可以用来制造氧气,氧气是制造燃料所需的氧化剂。
物 理	再水化的酱汁和炖菜由于表面张力而黏在一起;马尔顿海盐,这种片状海盐因其酥脆的口感以及对烘焙食品的良好附着力而受到人们的喜爱;晶体结构的改变会改变药物的疗效;火星的大气层比地球薄100倍;平均温度为零下60摄氏度。
工程学	在重力作用下,对流影响晶体的形成,所以在太空中容易产生几乎无瑕疵的大晶体。
经济学	一瓶水在飞船上要花1万美元,一只柠檬要花2000美元,这还只是计算携带它的能源成本;月球上的碎片,每克售价25000美元,是黄金价格的2500倍;每吨有效载荷需要7吨燃料;在太空中,我们不按千克购买食物,我们需要按营养价值衡量食物的价值。
伦理学	太空旅行和人造卫星造成了大量碎片,污染了太空空间,并将很快危及飞行,引发碰撞;在小行星上采矿会减少地球上采矿的压力。
历 史	1961年4月12日,苏联进行了第一次载人航天飞行,只绕地球运行了一圈;第一次成功飞越火星是在1965年7月14日至15日;第一次进入火星轨道是在1971年11月14日;陨石有许多秘密:比地球岩石要古老得多,这些来自太空的有趣的岩块可能保存了早于行星的矿物成分。
地 理	火星距离地球约5500万千米,距离太阳2.28亿千米;风笛最早出现在安纳托利亚(土耳其)。
数 学	无线电通信的延时计算;计算火星上的重力。
生活方式	我们的厨房用具都不能在太空中使用,这让我们重新思考饮食文化;我们每周改进菜单,使我们的食物摄入有规律;我们应该吃季节性食物,因为它能提供我们在冬季或夏季所需的营养;我们吃的新鲜食品越来越少,加工食品提供的营养越来越少,但保质期却越来越长。
社会学	人们在太空中不仅想要食物,还想要与地球上的饮食有关的仪式;食物反映了我们生活的生态系统;新鲜水果和蔬菜已成为一种奢侈品。
心理学	孤独的宇航员容易患上唯我论综合征,总览效应:一种合二为一的感觉,以及对地球美丽和脆弱的强烈感知,孤寂感(地球从视野中消失时产生的分离感)会影响未来宇航员的精神状态;缺乏正常的满足来源;在太空中的倦怠状态;太空适应综合征。
系统论	在宇宙飞船上,93%的水是通过最先进的液体回收系统回收的,该系统可以处理灰水、尿液、汗液,甚至可以吸收机舱空气中的水分;太空旅行的解决方案可以转化为地球上资源浪费的解决方案。

情感智慧
Emotional Intelligence

安妮

安妮显得很自信，认为能源问题属于已解决的问题。她明确了解决问题的首要任务：食物。令她担忧的是在一个狭小的空间里会变得抑郁。她知道人类以前也遇到过这个问题。她懂得变通，愿意从文化和态度上改变。考虑到艺术、运动和美食的需求可以满足，这使她下定决心离开地球。她善于观察细节，理解团队和日常事务的需要。她对于自己能够在太空中做一些在地球上永远不可能做的事感到兴奋。她坚信且不掩饰自己的想法，并决心解决讨论的问题，即使她知道父母可能不会同意。她在讲话时始终宽和待人。

凯瑟琳

凯瑟琳同样态度坚定。虽然没有细谈，但她提出了小行星开采这个有待解决的大问题，她很务实，希望关注尚未解决的问题。她倾向于长远的解决方案，并不太关心长达9个月的火星之旅。相反，她担心会变得无聊，并向安妮明确表示，这才是最大的挑战。凯瑟琳所渴望的并不仅仅是适应，而是创造一种新的生活方式。她畅所欲言，分享新奇的想法。她开放的思维也激发了安妮的强烈赞同。

艺术
The Arts

谈到餐厅，是时候跳出常规思维了。让我们想象一下，在一个与我们现在的逻辑和文化毫无关系的餐厅里用餐。盘子、餐具和椅子都不见了。如我们所知的厨房也不见了。这个创造性的练习必须说明在赛博空间是如何准备并享用食物的，以及如何确保食物不会进入你的鼻孔，而是进入你的嘴巴！

思维拓展
Systems: Making the Connections

我们现在的生活方式是不可持续的。种植、加工、分配、消费食物的方式让地球伤痕累累。世界上很多人都没有得到足够的健康食品。计划一次长达9个月的火星之旅，以及旅程结束后火星上的生活，为此要想出的解决方案显然是不同寻常的。如果我们能策划一场太空旅行，那么也能为地球生活找到可持续之道。前提是任何食物的成本都必须用宇宙飞船运送来计算。考虑新的计算方法：与其担心每千克的成本，不如参考单位营养成分的价值。如今人们消耗的许多食品不仅没必要还会导致肥胖和糖尿病等疾病。我们不仅要把重点放在食品和营养的技术上，而且所有社会成员都要保持变通，抛弃旧习惯，创造新习惯。我们在追求更生态的社会进程中面临的最大挑战可能是：准备好接受新的生活方式和饮食了吗？这就是太空旅行提供的转型机会。星球旅行需要团队成员相互信任，协同工作，在危机时刻保持坚韧。这是成功的先决条件之一：一个相信能够成功移居火星并在那里成功建立社区的团体。它将激励后来者。

动手能力
Capacity to Implement

准备一顿太空晚餐。虽然它不可能真的出现在零重力的太空环境里，但想象一顿没有盘子或餐具的晚餐。每样东西都是怎么准备和烹饪的。这是一个创意练习，而且很有必要。首先，所有东西都会物尽其用，不会浪费。所以如果你用香蕉，你会怎么处理香蕉皮呢？从新鲜水果开始，想一想如何处理种子和果皮。这样做的挑战在于，无论你做什么都不能有一点废弃物。接下来，考虑一下如何在不使用盘子的情况下吃到菜。这样的好处是不用洗盘子了！

故事灵感来自

This Fable Is Inspired by

艾利尔·埃克布劳
Ariel Ekblaw

艾利尔·埃克布劳是麻省理工学院媒体实验室太空探索计划的创始人和负责人，该计划由 50 多名研究生、教师和工作人员组成，他们通过 40 多个内部研究项目积极地构建科幻太空未来。在这项计划中，艾利尔负责协调太空研究和发射机会，横跨科学、工程、艺术和设计等领域，并在这项工作上与麻省理工学院和航天工业伙伴建立了合作关系。艾利尔同时也是麻省理工学院媒体实验室的助理研究员，她正在约瑟夫·帕拉迪索博士的响应环境小组攻读航空航天结构博士学位。她目前的研究包括设计、测试和部署自组装空间建筑，作为未来太空游客的栖息地，以及在环绕地球和火星的轨道上执行科学任务。

图书在版编目（CIP）数据

冈特生态童书.第七辑：全36册：汉英对照 /
（比）冈特·鲍利著；（哥伦）凯瑟琳娜·巴赫绘；
何家振等译.—上海：上海远东出版社，2020
ISBN 978-7-5476-1671-0

Ⅰ.①冈… Ⅱ.①冈… ②凯… ③何… Ⅲ.①生态
环境–环境保护–儿童读物—汉英 Ⅳ.①X171.1-49

中国版本图书馆CIP数据核字（2020）第236911号

策　　划	张　蓉
责任编辑	程云琦
助理编辑	刘思敏
封面设计	魏　来　李　廉

冈特生态童书
太空美食

[比]冈特·鲍利　著
[哥伦]凯瑟琳娜·巴赫　绘

颜莹莹　译

记得要和身边的小朋友分享环保知识哦！
八喜冰淇淋祝你成为环保小使者！